THE CRISIS OF THE TIGER

IS THERE A WAY OUT?

Shahbaz Ahmad

PublishAmerica
Baltimore

First printing

ISBN: 978-1-4489-8843-3
PUBLISHED BY PUBLISHAMERICA, LLLP
www.publishamerica.com
Baltimore

Printed in the United States of America

FOREWORD

Panna tiger reserve is currently in the news for the wrong reasons. It had reached a zero level for tigers and then relocation was taken up to rectify matters, but does it spell success? The reasons are now being sought by committees appointed by the Central and the State Governments.

At this juncture, Shahbaz Ahmad's title 'The Crisis of the Tiger—Is there a Way Out?' seems a welcome read. The author traces the history of tigers in the region and has a good chapter on the ecological basis of protection. The case history of the two "Bukkha' tigers is fascinating in respect of how tigers meet their end in nature, but alas! we are now mourning their unnatural deaths.

This small book appears to be special in two ways. First, the author has critically analyzed the theory of evolution in a very interesting manner. Secondly, he has developed the concept of *'Package of Principles'*, i.e. a set of strategies which have to be adopted simultaneously for the conservation of natural ecosystems. His suggestions deserve to be given due importance in Planning and Policy making.

Congratulations to Shahbaz Ahmad for his effort.

Date: 7-9-2009 **(J.J. Dutta)**
Ex Principal Chief Conservator of Forests
Madhya Pradesh, INDIA

Table of Contents

List of Figures 9
Preface 15

Chapter I: THE TURNING POINT 17
Chapter II: ECOSYSTEM AT WORK 21
Chapter III: THE ADVENT OF BUKKHA 30
Chapter IV: DEFENDING A TERRITORY 58
Chapter V: AN ERA OF CONTROVERSIES 60
Chapter VI: THE CAMOUFLAGED THREATS
TO NATURAL ECOSYSTEMS 62
Chapter VII: THE WAY OUT 66
Chapter VIII: EPILOGUE 76

REFERENCES 79
INDEX 81
Endnotes 86

LIST OF FIGURES

S. No.	Fig. No.	Description
1	Fig. 1.1	'Phansia' or trapper at Kehrvan near Panna
2	Fig. 1.2	Maharaja Laxman Singh of Dongargarh shot a tiger in 1940 near Baghai
3	Fig.1.3	Lokendra Singh, at the age of 15, killed a family of three tigers at Patori in 1962.
4	Fig. 1.4	The bench type topography of Panna National Park
5	Fig. 1.5	River Ken at Sankro
6	Fig. 1.6	The round the clock Nigrani (watch) and reporting to headquarters every hour
7	Fig. 2.1	A pair of Brahminy ducks
8	Fig. 2.2	The Dung Beetles, committed to their job
9	Fig. 2.3	The always busy Honey bee
10	Fig. 2.4	Indian Rock Python with a prey
11	Fig. 2.5	The Thoughtful Bukkha
12	Fig. 3.1	Bukkha without the left lower canine
13	Fig. 3.2	The reappearance of left lower canine in Bukkha
14	Fig. 4.1	The wounded Bukkha trapped by the camera
15	Fig. 4.2	125, a few days after the fight
16	Fig. 5.1	The seven tigers of Panna, caught by the camera
17	Fig. 6.1	Consequences of ad-hoc development
18	Fig. 7.1	Protected Areas are the producer of Water
19	Fig. 7.2	The sample collection of Pasai, the wild variety of rice
20	Fig. 7.3	Production Sectors in a Tiger Landscape
21	Fig. 7.4	The Package of Principles
22	Fig. 7.5	The Retreating Tiger
23	Fig. 8.1	The helicopter with the tigress from Kanha
24	Fig. 8.2	The Trophy of the Lion killed in Panna in 1929
25	Fig. 8.3	From an original oil painting by P. B. Fairbrother

ACKNOWLEDGEMENTS

My tenure in Panna was a rather short one—one year and nine months from June 2005 to March 2007. Yet there have been certain achievements worth mentioning. In spite of all odds, Panna could bag the award of the 'Best Managed Tourist Friendly National Park' of the country. The wildlife estimation exercise by WII in 2006 indicated a healthy tiger population in the Dry Deciduous Forests of Panna. The revelation of this fact gagged many vehement critics who had been bent upon defaming Panna. For the first time, the staff of Panna dared to have a face to face with the dreaded dacoits. A round the clock *Nigrani*(watch) system not only helped the Park management to have an eye over the forests and wildlife, but it could many times reveal the secretive movements of dacoits and trespassers.

All this could be achieved due to a team work with a dedicated staff. I would like to express my first gratitude to the forest guards, foresters, deputy rangers, the patrolling parties, the persons on the Nigrani towers, the mahawats and the characutters, the drivers and all the field staff who consistently toiled and performed their duties in a committed manner. Along with the field staff, the office staff, the wireless operators, the peons etc. also deserve thanks for their hard work and delayed sittings in the office. The Range Officers of Panna Tiger Reserve have been the principal executives to get the concepts and directions percolated and translated into action. They have played a major role in the establishment of the Panna Nigrani System and its implementation. The Assistant Directors Shri N. K.

Sanodia, Shri A. K. Parashar and Shri A. K. Kolte have been efficient bonds between the office of the Field Director and the field staff. They have acted efficiently in times of crisis and have played commendable supervisory roles. Also worth mentioning is Dr. S. K. Gupta, the veterinarian of Panna who is doing committed and selfless service to the wild animals. At times he has risked his own life to save the animal.

Shri A. K. Nagar, the Deputy Director requires special mention. In the tiger reserves of Madhya Pradesh much depends on the aptitude of the Dy. Director. It is because of the committed and disciplined attitude of Shri A. K. Nagar that decisions could be made quickly and things moved swiftly. With Shri A. K. Nagar as my colleague the team work was easy and challenges could be tackled boldly and efficiently.

Managing a tiger reserve means active interaction with the administration and police authorities. The Collector, Supdt. of Police and the other revenue and police officers of the districts of Panna and Chhatarpur have been helpful in many developmental and protection matters. The district Tiger cells have helped a lot in dealing with the dacoits and antisocial elements.

The one name in Panna which needs special mention is that of the founder of Panna National Park Shri Lokendra Singh ex-MP, ex-MLA and presently a member of the State Wildlife Board. He is a conservationist by heart and is a source of inspiration to the Park management. Koustubh Sharma is another person who has special affinity for Panna He worked on a Chausingha project in Panna and is now the Regional Co-ordinator for many countries on a snow leopard project. He is always ready to render his services to the Park management. Shyamendra Singh *alias* Binni Raja has special attachment to Panna. Professionally in the business of tourism he is also a member of State Wildlife Board and advocates strict conservation so that there may be rich wildlife in the Park. My thanks also to the present Field Director of Panna Tiger Reserve, Mr. R. Sriniwas Murthy and Deputy Director Mr. Vikram Parihar for providing all possible help in the collection of relevant information.

Any description about Panna National Park cannot be complete without the mention of the special support I received from the media. At times it appeared that the media was also a part of our management team. Many media persons became staunch wildlife lovers. Shri Shiv Kumar Tewari of Sahara Samay was so enchanted with wildlife that he himself shot many documentaries on Panna. Shri Arun Singh of Nav Bharat has a lucid style of writing. With his literary bent he projected the Park in a very responsible manner. Shri Shiv Kumar and Shri Arun Singh were always active participants of the different activities and campaigns which were conducted by the Park. Others who need mention are Shri Manish Mishra of Dainik Bhaskar, Shri Syed Javed Ali and Shri Nitendra Gurudev both of ETV. I owe my gratitude to all the media persons who have been will wishers of the Park.

Coming to the headquartes at Bhopal, my special gratitude to Dr. P. B. Gangpopadhyay, Additional Director General of Forests, Govt. of India who then as PCCF (Wildlife) was a guardian and guide to me. It was with his special efforts that the tiger crisis could be tackled efficiently and Panna could counter many false allegations. My thanks to Dr. H. S. Pabla, PCCF (Working Plan) who was then APCCF(Wildlife). He has been like an elder brother and always encouraged to take bold decisions. With Shri Suhas Kumar then CCF(Wildlife), I was more informal. In times of crisis he could give many valuable tips. Shri Ashok Bhatia CCF has been a helpful colleague while I was in Panna and also during my tenure as CCF(Wildlife) at the headquarters.

I also express my special gratitude to Shri Mohd. Hashim, PCCF and Shri V. R. Khare, PCCF. Both these respected officers (now retired) have been very compassionate and supportive in all diffcult times. My gratitude also to Shri Avani Vaish, then Principal Secretary, Forest who gave convincing responses to every request. My gratitude also to Shri P. K. Sen, former Director, Project Tiger and Shri Ravi Singh, Secretary General WWF-India. They visited Panna after I had left, yet they took the trouble to complement me for establishing the Panna Nigrani System. My thanks also to Dr. Rajesh Gopal, Member Secretary, National Tiger Conservation Authority, New Delhi who has been a moral booster for us. He visited Panna a number of times and lent all possible support to our efforts.

My special thanks to the highly respected Shri J. J. Dutta, former PCCF, MP and a wildlife expert of repute, who has accepted my request and has taken the trouble to write a foreword to this book. This will certainly enhance the value of this small effort.

The writing of a book requires much of concentration and a strict time management. It could be possible due to the strong moral support from the members of my family. They also deserve due gratitude. Bringing a manuscript in its final form requires frequent contributions from the personal staff. Hence I would specially thank Shri P. K. Dwivedi, Steno, Shri Vishnu Jadiya, Shri Vikas Shrivastava and Shri Naveen Kumar all of Panna Tiger Reserve and Shri Balram, Steno alongwith Shri Ashok Khare, Asstt. Programmer of MP MFP Federation where I am presently posted. My thanks also to Imam Mumtaz of Mumtaz Computers and my son Zeeshan Abdul Hannan. Special thanks to Dr. Vinita Chawdhry for giving her technical touches to the manuscript.

In the end, I would like to express my thanks to PublishAmerica USA for agreeing to publish the manuscript and bringing the book in its present form.

BHOPAL

11th Sept. 2009

Shahbaz Ahmad

PREFACE

The All India Wildlife Monitoring Exercise conducted by the Wildlife Institute of India in 2006 has indicated a drop in the number of tigers as compared to earlier claims. It means that the Natural habitats all over the country are under stress. To call it a crisis of the Tiger only will not be judicious. In fact it is a crisis of the Natural ecosystems—the life giving webs and networks. The crisis is more about human beings than the tigers. Yet there appears to be a marked complacency about the dwindling ecosystems.

It is here that the wild lifers and environmentalists appear to have failed. They have not been able to build up a focus on essential issues. For healthy ecosystems and real human welfare certain conscious decisions have to be made. Things cannot be left at the mercy of highly commercial market forces. Instead of economic indicators, eco-compatibility and resource sufficiency need to be denoted as positive indicators of progress and development.

This is a learning to which many have arrived at by a keen observation of Nature. In this aspect Panna National Park has been a great teacher to me. I have tried to ponder deep into the social, economic and ecological intricacies. I have tried to arrive at the real crisis behind the Crisis of the Tiger. I have tried to find a possible Way out of this crisis. And this I have tried to pen down in this little book.

I hope this book will be liked by all Nature lovers and it may be able to contribute something to the cause of the Tiger and Ecosystem Conservation.

BHOPAL
11th Sept. 2009
Shahbaz Ahmad

Chapter I
THE TURNING POINT

The district of Panna lies in the north western part of the State of Madhya Pradesh in India. It lies south of the district of Banda of the State of Uttar Pradesh. 'Panna' literally means emerald—a gem with greenish hue. The forests here are also known as the emerald forests due to this greenish hue. Panna is also said to have derived its name from the term *'Parna'*—the military camp of the Bundelas.

Panna was originally a Gond settlement upto 13th century. It was raised to importance by Maharaja Chhatarasal Bundela who made it his Capital in 1707. Maharaja Chhatarasal was able to free the greater part of Bundelkhand from the Mughals bounded by the Yamuna in the North, Narmada in the South, Chambal in the West and the Tons in the East. (Luard, C. E.: 1907)

The former princely State enforced a rigid ban on the shooting by the general public and the exclusive privilege of shooting was vested only with the ruling princes and their kiths and kin. It is evident that tigers were abundant in this region. Five km. from Panna is a place called Kehrvan which means tiger forest (Kehr tiger+van forest). Tigers roamed all over the forests in Panna. One of the Bundela kings Maharaja Nirpat Singh(ruled 1849—1870) was very fond of tigers and he had developed a device to catch live tigers. This device was known as 'Phansia' or trapper (**Fig. 1.1**). It consisted of a stone platform of about 3m in diameter with a long conical base of 10-14 ft height. It was surrounded by a tapering wall about 2m away with an opening at one side. A buffalo used to be tied on the platform and the gap around was covered with sticks and grasses so as to look like a plane ground. When a tiger tried to kill the buffalo it fell into the gap and could not climb up. It had no option but to move into the cage already placed at the opening and thus was captured alive. Raja Nirpat Singh had also developed a zoo and a garden. He is said to sit on the throne with tigers, trained to sit on his sides. But this affinity proved fatal to him. He was killed by a tiger in 1870 when he was out for shooting.

There would have been not so much concern for the tigers if the rulers had gone just on occasional shootings. But killing a tiger had become a symbol of valour. Kings and princes had become proud of the number of tigers they had shot. As if this was not enough, rulers of other princely States were invited to shoot tigers (**Fig. 1.2**). This hunting spree continued. Even young prince and princess took pride in being photographed with a tiger they shot. One such shooter was Lokendra Singh, the grandson of Maharaja Yadvendra Singh, the last ruler of the Panna State. Lokendra Singh shot his first tiger in 1955 when he was just nine years of age. In 1960, when he was 15 years of age, he shot a family of three tigers at a place called Patori in Chhatarpur district, which was later to become part of Panna National Park. Lokendra Singh sat proudly to get photographed with the three tigers (**Fig. 1.3**). But this was to be the turning point.

Something hurt the heart of Lokendra Singh, and he converted from being a hunter to become a tiger conservationist. He did not shoot any more. He was a member of the State Legislature from 1977-1980, became a member of Parliament in 1989 and was again MLA in 1995. He is also the founder Chairman of Panna National Park Formation Committee. It is with his efforts that we also have Panna Sanctuary and Ken Ghadiyal Sanctuary in Panna and Chhatarpur districts. But now the times have changed. Protected areas are not being looked upon as essential inviolable ecosystems which are in continuous service to the people. With a short sighted definition of development, protected areas are being targeted as hindrances to the so-called development. Lokendra Singh also bears the brunt of this faulty concept of development. He is often pin-pointed as the person who has halted the development of Panna by forming Panna National Park.

Concrete efforts towards wildlife conservation were initiated in 1975 when the Government of Madhya Pradesh notified its intention, under

Wildlife Protection Act, 1972 to form Gangau Sanctuary over an area of 285.337 sq. km. This proposed sanctuary was extended to 478.89 sq. km. and was renamed as Panna Sanctuary in 1979. It was in 1981 that superseding the previous notification, the government notified its intention to form Panna National Park over an area of 542.66 sq. km. It incorporated most of the area of erstwhile Panna sanctuary and also added a sizable forest area of Chhatarpur district. About 2/3rd of the Park lies in Panna district and about 1/3rd in Chhatarpur district. Ken river, the life line of the Park and the natural boundary between the two districts, meanders for 55 km in the Park. In 1984, the erstwhile Panna Sanctuary was revived and so we have four patches of the Sanctuary surrounding the Park and spreading over an area of about 78 sq. km. Being an important tiger habitat, Panna National Park was declared a Tiger Reserve in 1994 by the Government of India.

The forests of Panna broadly fall under the dry deciduous type. Yet the beauty and diversity of the Park are marvellous. The uniqueness of the Park is due to its bench type topography. At the top we have the Talgaon plateau followed by Hinouta plateau and then the Madla plains. This bench type topography has endowed the Park with a number of gorges and escarpments (**Fig. 1.4**). It is due to these that with the onset of monsoon we have a number of picturesque waterfalls. The emerald hue of the forests combined with the beauty of the falls and the flow of the Ken have an aesthetic appeal of their own (**Fig. 1.5**).

Panna is also unique in being the northernmost limit of Teak and the easternmost limit of Kardhai *(Anogeissus pendula)*. Acording to Champion and Seth's classification, we have the following six types of forests in Panna National Park (Chawdhry, P. K.: 2002) :—

Southern Tropical Dry Deciduous Dry Teak Forest—5A/C1b
Northern Tropical Dry Deciduous Mixed Forest—5B/C2
Dry Deciduous Scrub Forest—5/S1
Boswellia Forest—5/E2
Dry Bamboo Brakes—5/E9
Anogeissus pendula Forest—5/E1

At the international level Panna Tiger Reserve along with the adjoining forest areas falls in Panna Son-Ghadiyal Tiger Conservation Unit(TCU). It is a long strip running from Damoh to Sidhi district. It is a TCU requiring immediate surveys i.e. it potentially contains extensive blocks of appropriate tiger habitat, with or without protected core areas, but data on habitat quality, poaching pressure, or population status for the most important habitats within the TCU are lacking (Seidensticker, Christie & Johnson: 1999). During 2005-06, due to the special emphasis given to protection and other aspects of management, Panna could bag the National Award for the Best Managed Tourist Friendly National Park. A unique initiative was the introduction of the round the clock Nigrani(watch) system with reporting to the headquarters every hour. This has probably been the first and the only effort of its kind in any protected area of the country(**Fig. 1.6**).

Panna Tiger Reserve has an important role to play in tiger conservation. If the habitats in the whole of the TCU are managed properly and the connectivity of corridors is taken care of, Panna Tiger Reserve can serve as a rich breeding ground where tigers get nursed and then disperse into adjoining areas. It was in this protected forest that a tiger appeared in 2003 near Bhauradoh on the banks of Ken river. It was found to be lacking in one of its canines, so it was named ***Bukkha*** i.e. the one without teeth. We will meet him more intimately in Chapter III.

CHAPTER II
ECOSYSTEM AT WORK[1]

Strolling in a wilderness one often overlooks the dynamics of a natural ecosystem. If one is in a tiger reserve, the desire to see a tiger overshadows the subtler aspects of Nature. In a natural ecosystem there are miracles galore. One has only to develop an aptitude to see them.

On a sunny winter morning, if you are near a water body in a Protected Area, you are almost sure to see a number of birds, many of them migratory. One of these may be the orange brown Brahminy duck (**Fig. 2.1**). The nearest nesting ground of Brahminy duck is in Ladakh, Nepal and Tibet. The nesting season is April to June. During winter season it migrates southwards and is found all over India (Ali, Salim: 1996). The migration of birds is still a mystery. How is it, that the young ones follow their maiden migration on the same route as their parents! An often quoted example of the mystery of migration is the juvenile European Cuckoo, that is not only reared by foster parents of an alien species, but also embarks on its first long flight to southern Africa a month or more after its true parents have already departed. Yet, if you enquire about the details of migration, the ultimate answer will be that the birds have learnt it through evolution.

If you are a keen observer, you would sure have observed the dung beetle rolling the ball of dung with its hind legs (**Fig. 2.2**). Why is it involved in such a cumbersome process! The ultimate answer will again be that it has learnt to do it through ages of its struggle for existence.

If you are near a bee-hive in any protected area, your guide will certainly tell you to be away from it. Honey bees are social insects living in complex and very large colonies with a queen, the male drones and the sterile worker females (**Fig. 2.3**). With no hierarchical structure, it is a mystery how these different classes work in perfect harmony. Why does the male drone mate with the queen, just to die after this blissful event? Why does the worker female make a sting which is fatal to her? The most probable answer will be that these habits have evolved through natural selection.

If you are moving in a moist forest area during summer, you may often hear the song of the insect Cicada. Each cicada species has a unique song, usually produced by a pair of organs called 'tymbals' on the underside of the first abdominal segment (McGavin, George C.: 2005). The cicadas start singing all at once and then become silent all together. How they coordinate this is still a mystery. Why do cicadas take 4-17 years to become adult? The intelligent answer will again be that it is because of evolution.

If you are a frequent visitor to forest areas, you may have seen the Indian Python, either alone or with a prey inside its stomach (**Fig. 2.4**). How is it that the slow python is able to kill a swift deer in the darkness of the night? The answer will again be that natural selection has evolved such mastery in it.

These are but a few examples. Millions of such events go on daily in a living ecosystem. Every such event is a miracle in itself. Yet the miraculousness of Nature is overshadowed by the concept of Evolution. The strange thing is that even a lay person with no background of science, will explain with confidence the concept of evolution, and reduce the miraculous events of nature to an apparent comprehensible simplicity. People pretend to understand the complexity of ecosystem processes within the word 'Evolution'. If Evolution is such a 'magic-wand' which explains all the events of Nature, it would be worthwhile, to digress a little and ponder a while over this issue.

In the mid nineteenth century, Charles Darwin propounded the theory of Evolution. In his 'Origin of Species', Darwin has continuously tried to demonstrate that there is one general law leading to the advancement of all organic beings—namely, multiply, vary, let the strongest live and the weakest die. This is brought about by a power called Natural Selection, which is incessantly ready for action, and is as immeasurably superior to man's feeble efforts, as the works of Nature are to those of Art. Natural Selection has been defined as the preservation of favorable individual differences and varieties, and the destruction of those which are injurious. Darwin claimed that Nature can never take a great and sudden leap and that Natural selection will banish the belief of the continued creation of new organic beings, or of any great and sudden modification in their structures (Darwin, Charles).

All throughout Charles Darwin has tried to deny creation or a Creator. He has even gone to the extent of saying that—"I have nothing to do with the origin of mental powers, any more than I have with that of life itself." In fact Darwin was himself aware that, while putting forward his theory of Evolution, he was purposely avoiding certain issues. He emphasized that Natural selection has been the main but not the exclusive means of modification. Darwin never elaborated what these other means of modification were. As he proceeds towards the end of his treatise, he becomes more wavering—"I believe that animals are descended from at most only four or five progenitors, and plants from an equal or lesser number." He further proclaims—"...all the organic beings which have ever lived on this earth may be descended from some one primordial form." He has not elaborated what this primordial form could be. At last he accepts the existence of a Creator—

"There is grandeur in the view of life, with its several powers, having been breathed by the Creator into a few forms or into one."

But what is still more important is the caveats which Darwin had himself placed for his theory of Evolution. He very categorically states that—

> *If my theory be true, numberless intermediate varieties, linking closely together all the species of the same group, must assuredly have existed...evidence of their former existence could be found only amongst the fossil remains.*
>
> *If it could be demonstrated that any complex organ existed, which could not possibly have been formed by numerous, successive, slight modifications, my theory would absolutely breakdown.*
>
> *If it could be proved that any part of the structure of any one species had been formed for the exclusive good of another species, it would annihilate my theory.*

Darwin's theory of evolution lies refuted by his own caveats. No fossil remains have ever been found which could have been pointed as a missing link between species. When terrestrial strata and fossil records are examined, it is found that living creatures belonging to the Cambrian period(530-520 million years) emerged in the fossil record all of a sudden without any pre-existing ancestors. The absence of the transitional forms is not peculiar to the Cambrian period. Not a single transitional form verifying the alleged evolutionary "progression" of vertebrates—from fish to amphibians, reptiles, birds and mammals—has ever been found. Every living species appears in its current form, perfect and complete, in the fossil record (Yahya, Harun: 2002). The perfect harmony and recycling in Nature indicates that organisms are not just struggling for their own self-existence. They have their roles for the good of other species and the ecosystem as well.

The recent findings in the field of microbiology and genetics have made things more clear. Charles Darwin developed much of his theory by observing the breeding of domestic species. What was overlooked was the fact, that the breeders selected the traits which were already present. There was no production of any new trait which was preserved according to the theory of natural selection. With lack of genetic background, the morphological features were given prime importance. Thus it was thought logical to assume, that by continuous stretching of neck for thousands of years, a simple quadruped could turn into a giraffe, or a reptile, by continuous efforts to fly would turn into a bird. The recent findings show that the character of a living organism is already programmed in its genes. The genetic character cannot change by extraneous factors, even if tried by individual organisms for thousands of generations. Moreover, any genetic change, to be inherited, has to occur in a single life-time. Genetic mutations, wherever they occur, are only harmful to the species, and can't be put forward as a basis for evolution. Living organisms are on the whole irreducibly complex i.e. all its components have to exist at the same time. If any single component is missing or not properly working, the organism will die. Any modification in the irreducibly complex organism will only annihilate the organism (Yahya, Harun: 2004).

The theory of evolution has been unable to provide any explanation for the existence of the molecules that are the basis of the cell. The probability of the random formation of a protein and nucleic acid is inconceivably small. The chance against the emergence of even a particular protein chain are astronomic (Ali Demirsoy: 1984). In fact, it is not possible for the blind forces of nature to create even a single molecule of a living cell. To assume that living beings have developed and evolved by chance interactions can only be wishful thinking.

Today the theory of evolution lies refuted by its own caveats. Yet its initial effect was so profound that it continues to be accepted as a proven scientific fact. Everybody felt proud to be associated with Darwinism. People went to the extent of committing forgeries. In 1912, Charles Dawson, a well known paleonthropologist, claimed that he had found a jawbone and a cranial fragment in a pit in Piltdown, England. He asserted that the jawbone was more ape-like while the teeth and the skull were like a man's, thus displaying an absolute proof of human evolution. In 1953 the forgery was revealed through fluorine testing. It was found that the skull belonged to a 500-year old man, and the jaw bone belonged to a recently diseased ape.

In 1922, Henry Osborn, the Director of the American Museum of Natural History claimed to have found a fossil molar tooth belonging to the Pliocene period in western Nebraska. Based on this single tooth, reconstructions of man's head and body were drawn and there developed a "Nebraska man." In 1927 other parts of the skeleton were discovered and it was found that the tooth belonged neither to a man nor to an ape but to an extinct species of wild American pig.

Darwinism today exists more like a faith. Faced with so much of dilemma due to modern scientific discoveries, people try to justify the theory of evolution by proposing certain modifications to the original idea of Darwin. To explain the complex social behavior in certain animals, it is now being proposed that individuals will always be selected to sacrifice their own interest for the good of the group. It is now being accepted that without a significant modification of the theory of natural selection, altruism cannot be explained. It is being stated that apparent altruism at the level of the individual animal is no altruism at all at the level of genes; it is really selfishness from the point of view of genes (Gadagkar, Raghvendra: 2005). The justification for Darwinism appears to be more incomprehensible than the original theory of Darwin.

The eighteenth and the nineteenth centuries have been the period of mechanistic thinking. Selfishness of man was eulogized by Adam Smith in his "The Wealth of Nations." He advocates that good intentions may be less beneficial for the society than self interested intentions. The same self interest is reflected in the 'super-ego' of Freud, the 'class struggle' of Marx and Engels and the 'struggle for existence' of Darwin. It is only recently that this concept of rationality and selfishness been challenged by humanist economists (Lutz & Lux: 1988). It is now for the scientific community to investigate afresh into the origin of species and various aspects of natural history.

With this background, let us go again to the beginning of this Chapter and wander about in the same wilderness. We find that our Brahminy duck is not a simple product of blind natural forces. There are long-term, ultimate factors programmed in the bird's behavior that determine when they should arrive at their breeding grounds in spring or reach vital feeding grounds in autumn. A number of bird species have been shown to have a sun compass in their brain, and it is reasonable to suppose that this may exist in all birds. It has also been shown that birds have a star compass, a moon compass and a magnetic dip compass. It has been found that birds know how to reorient after they have been displaced and find their way back to their usual winter homes. The way that they reorient and find their way back to their destinations is the great unsolved mystery of migration (Burton, Robert: 1992). In this aspect, the brain of a bird is a miracle in itself and is much advanced than the human brain. Even the respiratory system of birds is unique to themselves and do not show any similarity with that of the reptiles from which the birds have been alleged to have evolved. Birds have a unique breathing system, in which the lungs are linked to three pairs of thin walled sacs. These sacs cause air to pass through the lungs in a single direction. The one way flow of air ensures more efficient extraction of oxygen and flushing of carbon dioxide, which allows birds to fly at great heights.

Birds are there not to struggle just for their own existence. They have important ecological roles to play. Besides helping in pollination, they are the Nature's controller of insect population. Without birds the world would be at the mercy of insects (Hamlyn: 1989).

If you sit down and observe the dung beetle, you will find that it forms a small ball out of dung or droppings of various animals. It lays its eggs in this ball, then rolls it to a safer place and buries it under the ground. When its larvae come out they feed on this dung (Invertebrates (Hindi); 2000). The dung beetle here is not just struggling for its existence. It is playing its untiring role in the recycling of nutrients. It is working not just for its own survival, but for the well being of other species also and the ecosystem as a whole. The dung beetle is contributing its might to the "zero waste" economy of Nature. In nature there is no waste. The ecological economics is much superior to the classical economics which is being followed all throughout the world. It is evident that human beings have learnt very little from Nature.

If you find yourself near a bee-hive, the first reaction would probably be to get away from that place, lest you get a bee-sting. The sting of a worker bee is armed with barbs pointing away from its tip, so that firmly lodged in the victim's skin, it cannot be withdrawn. When the worker bee attempts to fly away after stinging, the sting, the poison gland, and a part of her digestive system are torn away and left attached to the victim. For the worker bee, it is an act of suicide to protect the colony. Another very interesting thing about the honey bee is that the queens are known to mate with 10-20 different males who make a risky mating flight only once in their life-time. The queen can store sperm received from their mates in her spermathacae. They have accessory glands that supply nutrients to the sperm and keep them viable. The queen is perfectly capable of choosing the sex of her offspring. When she wants to produce daughters, she lets some sperm flow from the spermathacae into the oviduct, so that fertilized eggs are laid. The fertilized eggs develop into females, while the unfertilized eggs develop into males (Gadagkar, Raghvendra: 2005). The reproductive system of the honey bee appears to be more advanced than the human reproductive system. Each aspect of this tiny creature is an eye-opener. It is not a product of blind forces of nature, but well-programmed and perfected to perform its role in nature. Besides producing honey, it is one of the important pollinators of various crops.

When one hears a cicada sing, one usually ignores it as a noise of some small insect much below in the professed ladder of evolution. But if one tries to know more about it, one finds that its nymphs burrow into the soil after hatching, using their strong front legs. The nymphs live underground on roots of trees and take 4-17 years to reach adulthood, crawling up trees for their final moult. The role of cicadas in the working of ecosystems is to be further studied. But its peculiar life style vindicates that the stages of its life are not affected by extraneous factors, but are already perfectly programmed in advance.

One is often puzzled to see a slow moving python kill a swift animal like a deer. It becomes possible because of specialized heat sensing organs. Special heat-sensitive pits are found in the boas, pythons and pit vipers. Experiments have shown that at least some species of vipers are sensitive to changes in temperature of as little as 0.001°C. Using these organs, the location of prey can be assessed, even in total darkness. This implies that messages obtained by each pit can be compared and that the snakes can use this information to judge not only the position of the prey but also its range in the same way that animals with binocular vision can judge distances (Mattison, Chris: 1998). The infra-red sensory organs of snakes are one of the many sophistications of Nature which cannot be assumed to be developed by the action of the blind forces of nature.

If now a tourist moves in a National park or a Sanctuary with an unprejudiced mind, he will find miracles all over. Each plant and animal will be presenting itself to him with a uniqueness and perfection. As he ponders more, he discovers more beauty and sophistication. The concept of primitive and advanced organisms vanishes. Instead of a selfish struggle, he finds a purpose and a role for each living being. An insect appears to be as important as the tiger. A small insect is as complex and perfect as a large mammal. The design and complexity in both is not totally comprehensible, because there is always a lot to be learnt and discovered. It gradually dawns that this superb, awesome design and harmony cannot be evolved and developed by the blind forces of nature. It has to be created by someone who has limitless knowledge and for whom complexity is not a barrier i.e. for whom the creation of a fly is as simple as the creation of a human being. But this fact is hard to digest, because this would amount to the acceptance of an All-Knowledgeable and All-Powerful Creator. For centuries, this reality has been dubbed as unscientific or out of the purview of science. Similar is the case in every field of science. The Big Bang theory of the creation of universe has been hailed as the greatest discovery of the last century. But the question still remains as to where the original cosmic egg came from. Obviously, out of nothing. So there is someone, Who creates out of nothing. But Stephen Hawking, author of 'A Brief History of time' says that—"...questions such as who set up the conditions for the big bang are not questions that science addresses'. So who does? (Editorial. New Delhi: The Times of India: 11th Oct. 2006).

In spite of the recent scientific findings, Darwinism continues to wield its influence. It is being taught in schools as a proven scientific fact. Any book on natural history will almost certainly form evolution as its base. Carnivores are believed to descend from one common ancestor—a small shrew like creature, living in trees and feeding on insects (Macdonald, David: 1992). Tigers are said to evolve about fifty million years ago from miacids—the first true carnivore, which crept around in the tree tops (Thapar, Valmik: 2006).

Returning to Bukkha we find that he is somewhat sensible. Standing in his protected habitat, he seems to be somewhat perturbed by the doubts which have been raised over the identity of his ancestors (**Fig. 2.5**). He appears to reiterate that—"I am a tiger; my ancestors were tigers; my descendents will be tigers. No Evolution please"

CHAPTER III
THE ADVENT OF *BUKKHA*

In Panna National Park tourism had started in 1991. But tiger shows i.e. the sighting of tigers from elephant backs started only in 2001. In a tourism zone the commonly seen tigers generally get popular by some name. Whenever a new tiger is spotted in a tourism zone it becomes a talk of the day. Such was the case with Bukkha also. Everybody there was keen to spot the new tiger.

Jhabru Ekka, the characutter(grass cutter) was one of the first persons to spot Bukkha at the river Ken in the summer of 2003. At that time it must have been in the prime of his age, eager to establish a territory. He must have done this soon after because Bhakku Ram, characutter could see him mating the same year. Prajapati, the guide remembers to have spotted Bukkha in 2004 at Magradabri on river Ken.

There is one interesting observation from Koustubh Sharma. Koustubh came to Panna in Nov. 2002 to work on a Chausingha project sponsored by Bombay Natural History Society(BNHS). He is a keen wildlife observer and is among those who noticed the appearance of Bukkha in 2003. At that time one of Bukkha's canines was missing (**Fig. 3.1**). But Koustubh observed that in June 2005 the same canine appeared to be there (**Fig. 3.2**). Bukkha must have lost his canine at a very young age, perhaps in a fight with another tiger or during a careless grip of some heavy prey. The lost canine seems to have been replaced by the growth of a new one. It implies that if a tiger loses its

प्राणः प्रजा अनु वस्ते पिता पुत्रमिव प्रियम।
प्राणो ह सर्वस्येश्वरो यच्च प्राणति यच्च न।।

Prana *clothes the creatures, as a father his dear son.* **Prana** *truly is the Lord of all that breathes and does not breathe.*

(Atharva Veda: XI:4.10)

وَاللّٰــهُ خَـلَـقَ كُلَّ دَآبَّةٍ مِّنْ مَّآءٍ ج فَمِـنْهُمْ مَّنْ
يَّـمْشِـىْ عَلٰى بَطْنِهٖ ج وَمِـنْهُـمْ مَّنْ يَّمْشِىْ عَلٰى
رِجْلَيْنِ ج وَمِنْهُمْ مَّنْ يَّمْشِىْ عَلٰى اَرْبَعٍ ط يَخْلُقُ
اللّٰهُ مَا يَشَآءُ ط اِنَّ اللّٰهَ عَلٰى كُلِّ شَىْءٍ قَدِيْرٌ ○

Allah *hath created every animal of water. Of them is (a kind) that goeth upon its belly and (a kind) that goeth upon two legs and (a kind) that goeth upon four.* **Allah** *createth what He will. Lo!* **Allah** *is able to do all things.*

(Al-Quran: 24.45)

Fig. 1.1 'Phansia' or trapper at Kehrvan near Panna
(Photo: Naveen Kumar)

Fig. 1.2

Maharaja Laxman Singh of Dongargarh shot a tiger in 1940 near Baghai (Courtesy: Lokendra Singh)

Fig. 1.3 Lokendra Singh, at the age of 15, killed a family of three tigers at Patori in 1962 (Courtesy: Lokendra Singh)

Fig. 1.4

The bench type topography of Panna National Park
(Courtesy: Koustubh Sharma)

Fig. 1.5

River Ken at Sankro (Photo: Koustubh Sharma)

Fig. 1.6

The round the clock Nigrani (watch) and reporting to headquarters every hour (Photo: A. K. Nagar)

Fig. 2.1

A pair of Brahminy ducks (Photo: Koustubh Sharma)

Fig. 2.2

The Dung Beetles, committed to their job (Photo: A. K. Nagar)

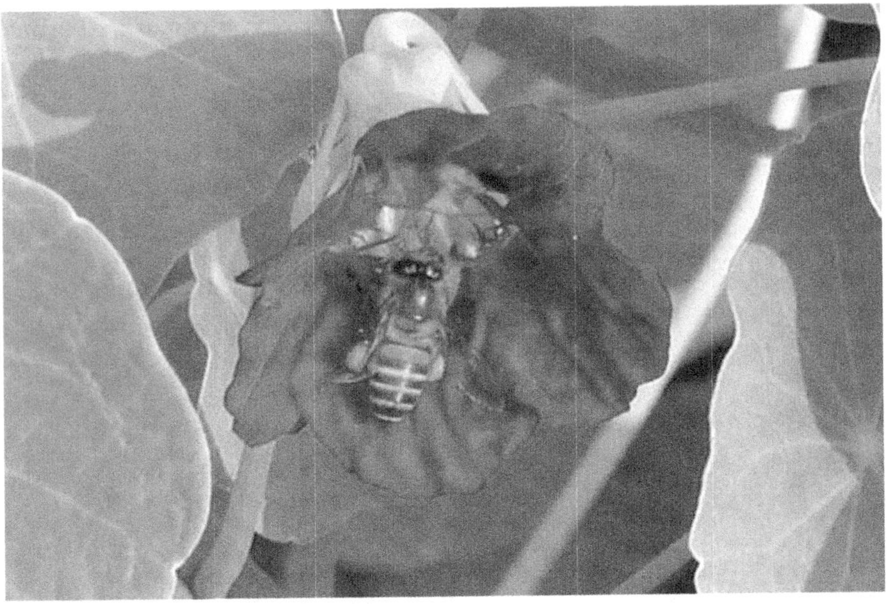

Fig. 2.3

The always busy Honey bee (Photo: A. K. Nagar)

Fig. 2.4

Indian Rock Python with a prey (Photo: A. K. Khare)

Fig. 2.5

The Thoughtful Bukkha (Photo: Koustubh Sharma)

Fig. 3.1

Bukkha without the left lower canine (Photo: Koustubh Sharma)

Fig. 3.2

The reappearance of left lower canine in Bukkha (Photo: Koustubh Sharma)

Fig. 4.1

The wounded Bukkha trapped by the camera

Fig. 4.2

125, a few days after the fight (Photo: A. K. Nagar

Fig. 5.1

The seven tigers of Panna, caught by the camera

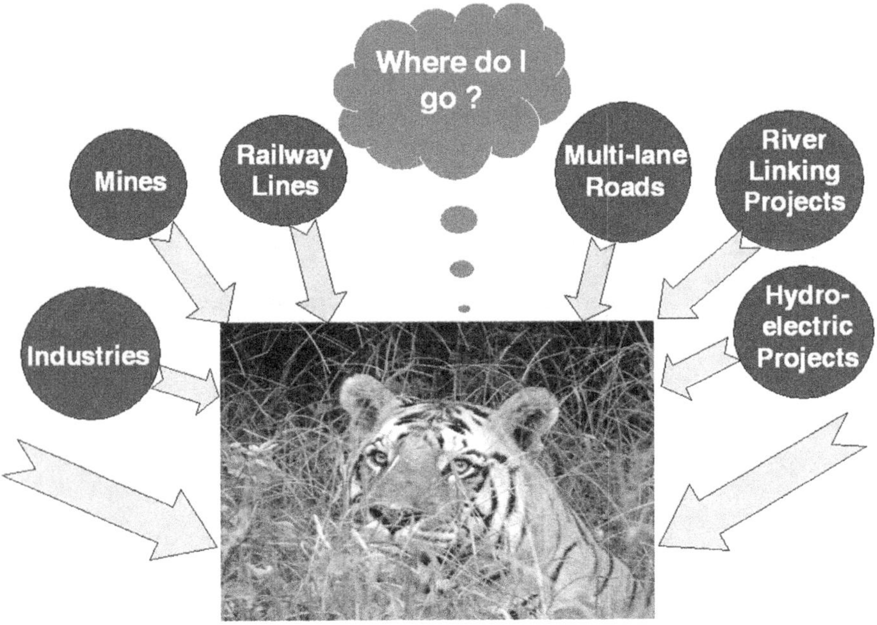

Fig. 6.1

Consequences of ad-hoc development

Fig. 7.1

Protected Areas are the producer of Water (Photo: S. Ahmad)

Fig. 7.2 The sample collection of *Pasai*, the wild variety of rice (Photo: A. K. Nagar)

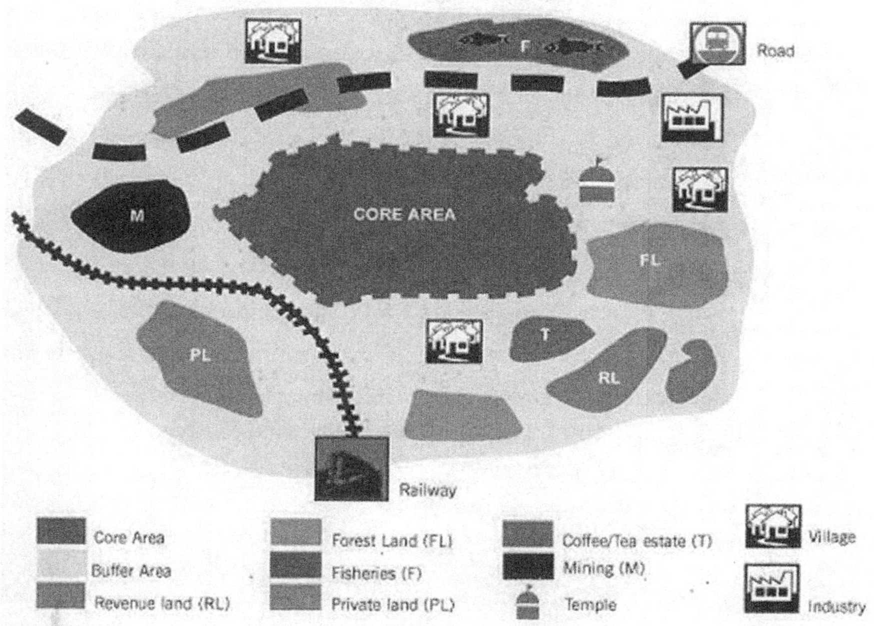

Fig. 7.3

Production Sectors in a Tiger Landscape (Courtesy: NTCA, Govt.of India)

Fig. 7.4

The Package of Principles (For sustainability of Ecosystems certain strategic principles will have to be adopted and implemented simultaneously)

Fig. 7.5

The Retreating Tiger (Photo: A. K. Nagar)

Fig. 8.1

The helicopter with the tigress from Kanha (Photo: A. K. Nagar)

Fig. 8.2
The Trophy of the Lion killed in Panna in 1929 (Photo: Arun Singh)

Fig. 8.3
The Relaxing Bukkha (From an original oil painting by P. B. Fairbrother)

teeth it is gradually replaced by the growth of a new one. This hypothesis needs to be verified by more observations.

In Panna National Park there is another tiger, senior to Bukkha who roams over a large portion of the tourism zone. It was radio-collared during a tiger ecology project. It was then known as 125, a name related to its radio frequency. Before that it was also known as Bukkha because it also had a missing canine. Because of this similarity there has often been confusion over the two Bukkhas.

Along with the senior Bukkha ten tigers were radio-collared in Panna during a decade. The apparent purpose of radio-collaring is research and study. Yet there is another side of the coin. For radio collaring a tiger has to go through a phase of tranquilization and then it has to continuously carry a sufficiently heavy weight on its neck. This might be hampering its routine activities and hectic pursuits. There might be some adverse effect on mating behavior also. In fact, the first radio-collared tigress in 1996 died due to strangulation when its radio-collar got entangled in a bamboo clump. Tranquilizations and radio collaring may be a source of continuous irritation to the animal and it might be giving distress signals to other members of its species. A study appears to be essential on the various effects of tranquilizations and radio-collaring, so that there is no torture of wild animals in the name of research.

With the appearance of the new Bukkha, a tussle for territory was inevitable. If both have to survive there has to be a division of areas. But this settlement can't happen peacefully. A fight between the two Bukkhas was noticed in March 2004.

Chapter IV
DEFENDING A TERRITORY

Territorial animals have an inherent tendency of marking and defending their territories. This behavior expresses itself differently in different species. Tigers are ferocious top carnivores, the males being more concerned with occupying large territories. This gives them a type of proprietary right over the tigresses and prey base in their area. But this territorialisation doesn't mean strict compartmentalization of the forest area or the formation of some kind of mosaic. The territories are often overlapping and direct confrontations are avoided. Why to have a fight if verbal threats can do the job? After all, a severe wrestling may prove to be fatal.

A well grown up tiger may for a few years enjoy the merits of its seniority. It is well known to other tigers who will often try to avoid quarrelling with their senior colleague. But at times it becomes necessary to show one's strength. An ageing tiger must win a wrestling bout to demonstrate that it has not gone weak. This may give him a respite for another few months. A sub-adult must also have a fight to show its budding energy and to carve out a territory from that of some other tiger. Same had been the case with Bukkha. Coming to age in 2003, it had to establish itself in a new area. The tourism zone of Panna National Park was to be the battle ground for the two Bukkhas.

There had already been a number of bouts earlier. But a real fight took place in the month of March 2006. On the 3rd of March a wireless message came that Bukkha was lying injured in Gariyadol nallah in Hinota range. We had to move quickly to the spot. The severity of fight could easily be guessed. There were wounds all over the body. Flies were hovering over these wounds. Bukkha tried to avoid the flies by covering his wounds in water as much as possible. But he seemed to be very weak. He could hardly walk a few steps. It appeared that he had been hungry for a long time. It was certain that he was not able to hunt. If left to himself he would have died.

A goat was sacrificed. Teams were organized which moved on elephants and threw big pieces of flesh and bones near him. Bukkha observed, stood up and walked, took a big piece in its mouth and walked away. When it started eating it was late evening and we also left the area to let Bukkha have a hearty meal. Next day it was reported that all the pieces had been devoured. A constant vigil was being kept on him. A few days later it was reported that Bukkha was moving with a tigress. Such is the recovery power of a tiger.

During the same period, a wild life estimation exercise was being conducted with the help of Wildlife Institute of India, Dehradun. Camera traps were fixed at a number of spots. Bukkha also got caught in the camera (**Fig. 4.1**). From the photograph it was clear that Bukkha was severely injured and his wounds were not completely healed up. The interesting thing was that although Bukkha was much younger than 125, it was he (Bukkha) who got most of the injury. A few days later 125 could also be spotted and photographed (**Fig. 4.2**). He had only minor injuries on his body. Perhaps experience had made the senior Bukkha a more skillful fighter.

Later two more male tigers frequently made their appearance in the tourism zone. Quarrels and fighting between these four males were often reported. These are just a few of the glimpses of the vast ecosystem activities continuously going on.

Chapter V
AN ERA OF CONTROVERSIES

The beginning of the 21st century has been an era of controversies for Panna. This happens often with a tiger country particularly that with a tourism zone. If tigers are not easily visible questions are raised about their numbers. A pugmark census in Jan. 2005 estimated the number of tigers in Panna to be 34. Some people were critical about the whole procedure. So a repeat census was conducted in March the same year. The number again came out to be 34. But many people were still not convinced.

There were criticisms and media reports that only a few tigers remained in Panna and that also only males. Public Interest Litigations were also filed in Supreme Court. The whole pugmark methodology was challenged.

In the meantime, the Project Tiger Directorate (later to become the National Tiger Conservation Authority), in collaboration with Wildlife Institute of India, came out with an improvised technique. This technique involved the statistical estimation of wild animals from a sample of evidences collected on linear transects. In tiger occupied areas, camera traps were also to be employed.

In Jan. 2006 the wildlife estimation exercise was conducted in the whole of Madhya Pradesh. In Panna the camera trap exercise was conducted from January to June 2006. Seven different tigers were trapped in cameras—4 males, 2 females and one with unidentified sex (**Fig. 5.1**). The photographs initially silenced the critics for sometime. Sadly enough, in May 2006, one tigress and her cub became the victim of poisoning. Their bodies were detected by patrolling parties deep inside the forest. Although the offenders were later caught and sent to jail, the incident outlined the need for more vigilance and caution. Soon there was another allegation—that there were only these seven tigers in the Park.

The Wildlife Institute of India analyzed the data for Panna. They reported that Panna has a healthy tiger density. The number of tigers estimated ranged from a minimum of 15 to a maximum of 32. Even this estimation could not satisfy the critics. They said that the average estimation of 24 is much less than the claimed population of 34.

Was the earlier census figure highly inflated? Or the improvised technique gives us a lower estimation? Which of the two estimation techniques is more precise? We may never be able to find satisfactory answer to these questions. Along with a strict protection regime, it is better to follow one methodology and to continuously monitor the changes occurring in tiger occupied areas and habitat availability.

While the allegations and controversies went on, the staff of Panna continued to strive sincerely to improve the protection and management of the Park. The efforts ultimately bore some fruits. Panna was adjudged as the **'Best Managed Tourist Friendly National Park'** of the country for the year 2006. The award was received on 25th January, 2007 in New Delhi by the then Field Director (the author of this book) from the Minister of Tourism, Govt. of India.

But there is no place for complacency in a wildlife rich area. Continuous vigilance and struggle has to be the motto. Besides all other efforts, the greater cause of concern is the biotic pressures on tiger habitats, resulting in their degradation and fragmentation. The precise number of tigers is not as important as the knowledge about tiger occupied territories. In the controversy over the number of tigers, the real issues of concern are generally overlooked. The real threats to tigers and natural habitats are often well camouflaged.

Chapter VI
THE CAMOUFLAGED THREATS TO NATURAL ECOSYSTEMS

Rumors and controversies are bound to be there in tiger areas. Bukkha is not so much perturbed from these as from the gradual realization that the local and global environment around him hasn't remained so friendly. He now feels more the pressure on his habitat. The rising human population is being seen all around. Instead of devising alternate solutions to meet their basic needs, people are only eying on his habitat. Everyday women could be seen violating his peace to collect fuel wood. With no pasture lands left, the only grazing ground available is his habitat. People can also be seen to collect other non-timber forest produce from the protected areas.

Legally these trespasses into Bukkha's habitat are not allowed and the Park management continuously strives to contain them. But checks and chasing alone can't solve the problem. Conservation and livelihoods issues have always been at loggerheads. There have been conservationists and the lobbies of human right activists, each trying to put their point of view as strongly as possible. In this aimless tussle, the problem is hijacked by some intelligent persons who profit in the name of 'people'. Appealing ideas, lucrative proposals and colorful presentations are often able to convince the decision makers about making drastic interferences in natural ecosystems. Short term gains are then preferred over long term and sustainable ecological benefits.

The intricate life web of ecosystems is lost sight of and wild life is reduced to a mere bunch of commodities which must be killed or utilized otherwise they don't have any value. A few influential persons and business houses then decide what is to be done to the ecosystems. The highly selfish commercial ventures are then run in the name of innocent terms e.g. resource farming, ecotourism, conservation hunting etc. The captive breeding ventures in USA and China are eye openers (**Box 6.1**). Although the purpose of the South China Tiger programme was professed to be the reintroduction of tigers in the wild, the two largest Tiger breeding centers in China are primarily commercial, both nearing or exceeding 1000 Tigers. Large, individual captive populations also pose a health hazard. In 2004, an outbreak of avian flu broke out at Thailand's Sri Racha Tiger Zoo. More than 140 tigers died and were killed by euthanasia by government disease control agents. Commercial ventures can't lead to conservation of ecosystems. (**Box 6.2**).

When commercial ventures and monetary profits lead the way, it ultimately leads to cruelty in a sophisticated manner. Some wealthy people derive pleasure from killing of animals. They do it just for fun and are proud to have a photograph with their trophy. Since they are able to pay a huge fee, they manage to modify the rules according to their convenience. Cruelty is masked by an apparent philanthropy. This cruelty is nurtured and bred and may be the possible root cause of many social evils.

These ventures do not stop here. Some showcases and eyewashes are created. It is said that this luxurious ecotourism ultimately helps the local people. But the real beneficiaries are others. Luxury hotels and airstrips are developed right inside the protected areas. The protected areas in Africa are large in size and may to some extent be able to tolerate these interferences. But it is strange that there are suggestions to emulate these practices in the already burdened and fragmented forests of India.

In India, hunting of wild animals is banned except in very special circumstances. Even the customary hunting of tribal is not allowed.[2] Yet, after a ban of a few decades, the hunting lobby seems to have gained some confidence. There is again a demand for organized hunting. It is being propagated that it will generate revenue and will be helpful to the people. It is also being claimed that this will be helpful in conservation. The basic question which arises is, "*Is the conservation of ecosystems dependent on the charity of some gun wielding millionaires who derive pleasure in killing?*"

Besides the profit oriented commercial inroads, the other major camouflaged threat to the natural ecosystems is in the name of 'development'. Their have to be roads, rails, bridges, aerodromes, irrigation, mining, industries etc. etc. But no one knows to what extent. So they go on increasing. Had it not been for the stringent laws in India, we couldn't have been able to save the existing forests and wildlife of the country. But now the situation appears to be in a decisive stage. The pressure of 'development' is increasing. The will of the ecosystems to survive appears to be fading. If development means meeting the basic needs of the people in an eco-compatible manner, this can certainly be achieved through a planned approach. But development has never been like this. "Development" again is fuelled by commercial forces. So, more development means more business. In this race of development, the real gainers are only a small percentage of the population. The livelihood issues of a vast population remain unattended. The forests and natural ecosystems are made scapegoats for this dilemma. The forests and wildlife are then classified as hindrances to development. A public opinion is generated which perceives the natural ecosystems as something to be got rid off. Many people find it convenient to go along with this opinion. To speak otherwise and to bring about the truth may require a lot of homework.

With all these developments going on, Bukkha feels himself cordoned off with no strong helping hand around him. (**Fig. 6.1**). He wonders—***"Is there a way out?"***

* * * *

BOX 6.1

In recent years captive breeding of Tigers in China has accelerated to the point where the captive population exceeds 4000 animals. Three thousand specimens are reportedly held by 10-20 "significant" facilities, with the remainder scattered among 200 facilities. This makes China home to the second largest captive Tiger population in the world, after the USA, which has an estimated 4692 captive Tigers, down from an estimated 7568 in 1997. These two countries have a marked difference in their approach to the breeding of Tigers in captivity. In the USA, captive breeding cats in private ownership outside accredited zoo breeding programs are viewed as a problem, and the government has passed legislation and is implementing regulations to reduce the size of the captive population. In China, private

Tiger ownership is growing rapidly, with national government support.

[Nowell, K. and Xu, Ling. (2007)]

* * * *

BOX 6.2

China's Tiger captive breeding facilities have done well in their goal to breed Tigers, but a costly and potentially dangerous surplus is now evident. China's Tiger captive breeding facilities have high running costs and are uneconomical, placing a financial drain on the State's Tiger conservation resources. Re-introduction into the wild of captive tigers from these facilities is not supported by the scientific community. The Tigers in China's captive breeding facilities are diverting resources and effort from conservation of wild Tigers and their habitat, and sale of captive-bred Tiger parts would compound the harm to Tiger conservation.

[Nowell, K. and Xu, Ling. (2007)]

Chapter VII
THE WAY OUT

Bukkha is sensing some danger. It means that the whole tiger habitat is at risk. This implies that the life giving ecosystems are under strain. Life and livelihood seem to stand face to face. There has to be a way out of this dilemma.

The way out is possible only if there is a holistic approach to the problem. If a natural ecosystem is looked upon in isolation, it appears to be no more than a problem for human beings. But with a sincere pondering, the horizons become clear and reality tries to exhibit itself.

Natural ecosystems are not just a conglomeration of plants, animals and other living beings. It is abuzz with life giving activities. They are producers of water—very essential for the existence of humans and other living beings (**Fig.7.1**). They are breeding grounds for many birds and insects, which maintain a biological control and help a lot in the pollination of plants and production of crops. They are a repository of plants useful in preventing and curing many diseases (**Fig. 7.2**). Gradually it dawns that large chunks of ecosystems need to be left untouched and preserved as such.

But then there are human beings all around. They need space, food, clothing and shelter, which have to be extracted from these ecosystems only. It is here that a strong caution is required. The urbanized centers start acting as pullers of the ecosystem products. The people in and around these ecosystems are used as tools to extract and transmit these products. The real beneficiaries turn out to be the urban centers and people living far away from these ecosystems. Instead of recognizing these urban pulls, which are backed by commercial forces, the impoverishment of the forest dwellers and rural people is ascribed to the existence of forests and natural ecosystems.

To tackle this situation, it is first of all essential to accept that a natural ecosystem is the best land use. Whatever management intervention is necessary, it should be driven by dire necessity and with a conscious acceptance that it is a worse option. This can be achieved, only if there is a long term landscape planning in which landscapes are allotted different uses and this is sacredly adhered to. But even this may not be sufficient. The 'rush-on' development will have to be replaced with an overall planned eco-compatible development. If there has to be a way out, everything which is not eco-compatible has to be phased out.

Eco-compatible development has to be a *package of principles* which have to be applied simultaneously. The agricultural production has to be organic, the basic needs have to be given preference over non-essential demands, industries have to recycle their wastes (which means there has to be a zero waste economy and only those industries are allowed to exist whose raw materials are obtained from non-polluting sources in a sustainable manner), herbal health care is given strong preference over other forms of drug therapies, etc. These decision making principles at first sight appear to be only remotely connected with the real issue. But the fact is that *the crisis over the tiger and its habitat is looming large just because the issue of eco-compatible development has been continuously neglected.* A Gandhian model of development appears to be the only way out to save the natural ecosystems (Ahmad, Shahbaz: 2002).

The package of principles for eco-compatible development may consist of the following approaches—

A Landscape Planning and Policy: The top priority of governments should be to adopt a landscape planning i.e. allot on a long term basis e.g. 100-200 years, uses to different portions of land. In a vast landscape there will be protected areas and other forests, railways, roads, agriculture, mining, aerodomes, residential areas etc. i.e. there will be everything essential for a civil society, but in a planned and committed manner. It is for the first time that the Government of India has strongly emphasized landscape planning for tiger landscapes (Fig. 7.3). The National Tiger Conservation Authority has issued guidelines regarding the preparation of Tiger Conservation Plan

(Guidelines for preparation of Tiger Conservation Plan—Technical Document: NTCA/01/07; 2007). For a proper landscape planning, the forests need to be classified according to the IUCN guidelines. Here the forests are classified as—

Strict Natural Reserve/Wilderness Area: Protected area managed mainly for science or wilderness protection.

National Park: Managed mainly for ecosystem protection and recreation.

Natural Monument: Managed mainly for conservation of specific natural features.

Habitat/Species Management Area: Managed mainly for conservation through management intervention.

Protected Landscape/Seascape: Managed mainly for Landscape/ Seascape conservation and recreation.

Managed Resource Protected Area: Managed mainly for the sustainable use of natural eco-system.

The IUCN classification will help in setting clear cut and long term objectives for the management of forests. People will know that different forests have been set apart for different uses. While Class I forests will be totally protected, other areas will be accessible to the people according to the classification and objectives. This may help in resolving out much of the conflict between the people and the forest department.

The landscape planning has to be both at the macro and micro levels. Previously every village used to have its own grazing ground and a small village or panchayat forest. These lands have gradually been encroached or transformed into residential or agricultural areas. The result is that the adjacent natural forests are being denuded at a very fast rate. It is time to realize this administrative neglect and to revive the concept of self reliant villages. If natural habitats are to survive, the concept of landscape planning at the village level also has to be pursued with a strong will. A village or a group of villages should be self reliant in matters of food, clothing, shelter, education and basic health care.

A Zero Waste Economy: The classical economics of Adam Smith is based on the false assumption that every human being is selfish. Hence all human beings are engaged in their selfish pursuits in one way or the other. It is claimed that the output of these hectic selfish pursuits is social welfare. Thus every activity which is profitable is justified in an economic sense. Hence an industrial venture is justified as long as it is economically profitable. It doesn't matter if its raw material is obtained in a manner which is detrimental to the environment. It doesn't matter if its output is a product which is not necessarily essential for humankind. It doesn't matter if it is causing pollution or producing a lot of waste. There is no place for ethics, principles or values in economics.

The economics which is in vogue is in sharp contrast to the economics of natural ecosystems. In nature living beings don't exist for themselves. They are engaged in functions which are beneficial to humans and other living beings. In nature there is no production of waste. The excretion of one living being is food for the other. The nutrients are always being recycled. If we have to conserve natural ecosystems we have to emulate their styles. The economics of human society should also tend to be a Zero Waste economy i.e. only those industrial units should be allowed to come up which have satisfactory processes for waste recycling and reutilization. Industrial units should not be allowed to produce waste and dump it as hazards for the whole living world.

The concept of zero waste economy has been purposely overlooked. The result has been the grave pollution of rivers, water sources and the environment as a whole. The grave consequences of this state are too evident to be mentioned. A recent example is the death of more than hundred gharials (*Gavialis gangeticus*) in river Chambal in a matter of just 2-3 months i.e. Dec. 2007 to Feb. 2008. The cause of death has been found to be kidney failure due to some nephrotoxicant. The nephrotoxicant is supposed to have come through fishes from the highly polluted Yamuna river.

Once the principles of zero waste economy is accepted and implemented, the ancillary benefits will themselves crop up. The emphasis on meeting basic needs will itself assume significance. Options for activities with higher employment generation—investment ratio will also increase. Any attempt to move in this direction will be beneficial both to natural ecosystems and to the people living in and around these areas.

Organic Farming: Agriculture is one of the many sectors where the multinational corporate lobby seems to have a strong hold. The use of synthetic fertilizers and pesticides has spread all over the world. In fact it is a well accepted evil and people have been made to believe that if man has to survive, the use of chemical fertilizers and pesticides is a must. These chemical products, especially the pesticides, are severely affecting the health of both the human beings and wild life. Organic farming is either derided or limited to demonstrative purposes. There is a big business behind the apparent efforts to increase crop production through synthetic and costly inputs. The reality is that after the initial spurts, the production gradually stagnates or goes down and the soil becomes barren. The high input cost can then be borne by big farmers only and the small and marginal ones have no option but to sell their so called unproductive land.

The reality has begun to dawn in some places. In Andhra Pradesh, in more than 400 ha, farmers have stopped using pesticides on cotton, without anybody asking them to do it. "Pesticides on cotton mean a guaranteed loss. It's suicidal. There is no difference in yield whether you use pesticides or not. If you do, the cost of cultivation becomes unaffordable," says a farmer. There is a movement in Andhra Pradesh called the non-pesticidal movement or NPM (**Box 7.1**).

In the interest of humankind it is essential that this awareness is spread far and wide and the enslaving tendencies of the corporate world are checked in time. The trend of using chemical fertilizers and pesticides has to be reversed if agriculture has to be sustainable and compatible with natural eco-systems. Organic agriculture has to be adopted as a policy so that the use of synthetic fertilizers and pesticides is minimal.

Chemical fertilizers and pesticides have done considerable damage to the natural biodiversity. The agricultural biodiversity is hardly visible. Every year hundreds of peacocks and other birds die due to intake of chemically treated seeds. Tigers are killed by poisoning the carcasses of their kills with chemical pesticides. The onslaught on Nature is taking its toll.

Herbal Healthcare: Healthcare is another field where pharmaceutical companies manage the affairs. The qualified doctors have no option but to prescribe what is produced and suggested by these companies. The health of a poor person in a remote village is dependent on some injection or tablet produced far off in a factory.

In India we boast of a tradition of Ayurveda and herbal health care. Yet this art and knowledge is hard to be seen in practice. What we see in the name of Ayurveda is again a business. In the field of health care, the true empowerment of the poor and villagers will be to enable them to grow and use the herbs directly for their health care and cure. Kalmegh(*Andrographis paniculata*) has long been known to be an efficient herb for the prevention and cure of malaria. Bhuiaonla(*Phyllanthus amarus*) is a herb commonly growing in rainy season, which is almost a specific for jaundice. Chakora(*Cassia tora*) is again a rainy season weed which has been found to be effective in flourosis (**Box 7.2**). There is no dearth of such examples.

In fact the tradition of herbal health care can easily be revived and people can be confident to use and grow herbs for their common ailments. But, here again the difficulty arises. The efficacy of herbal plants is questioned. Then the issue is hijacked by pharmaceutical people. They try to locate some 'active principle' in the herb. They then convert it into some injectable liquid or a pop able tablet. So a poor man is again at their mercy.

It is for governments to take steps to get rid from the clutches of allopathy and its side effects. It may be recalled that vultures have become almost extinct due to an allopathic drug 'Diclofenac' which was being extensively used in veterinary practice. The propagation of herbal health care will be a great service to human beings especially the poor and those living in remote and forested areas. It will also revive the emphasis which used to be given to different types of plants for their inherent usefulness in many ways. This will go a long way in reestablishing a symbiotic bond between the forests and the people.

Compensating damage by wild animals: In wild life rich areas, damage to human life and property is a common thing. People and domestic animals are killed or injured. The more severe problem is the crop raiding by herbivores and omnivores. With dwindling forest areas and emphasis on monoculture crops this problem has further accentuated. Since the farmers and villagers who suffer the losses are mostly the poor people, it is essential that a scheme of compensation is in place. There has to be adequate compensation for loss of human life and kill of livestock. In case of crop damage there has to be adequate compensation or subsidy. In the State of Chattisgarh, there is a provision for monetary compensation for crop damage by wild animals. In Gujarat there is a subsidy to farmers who erect a fence around the fields. In Madhya Pradesh efforts have been made to minimize crop damage by erecting crop protection wall and fences. Monetary compensation for crop damage by wild animals is also now in place. Whatever scheme is there, the system of compensation or subsidy should be reasonable enough to prevent the eruption of any strong antagonism against forests and wildlife.

No killing or torturing of wildlife for fun or revenue: In principle the forests and protected areas should be managed in such a way that the interventions produce minimal disturbances. Some revenue is also welcome if the interventions are well planned and they do not compromise with the overall naturalness of Nature. But all along it has to be accepted that natural ecosystems have their intrinsic values which are much more than any possible revenue earned. It is a mere mockery of Nature if we try to put any monetary value to its existence. If any wildlife is immediately threatening the life or property of human beings, the interest of human beings will certainly get a priority. If necessary, a number of wild animals may be killed, but with all sobriety. As a mark of respect to Nature, the least, which a civil society can do, is to pledge that it will not try to please the gun wielding killers. There has to be a complete ban on killing or torturing wildlife for fun or revenue. A dichotomy between the purpose and means should not exist.

The Eco-compatible development of villages: While it may be necessary to carry out relocation of certain villages with the voluntary consent and involvement of villagers, it is also essential to place special emphasis on meeting the basic needs of the people residing in and around PAs. These areas may be declared as eco-sensitive zones. It means that here no eco-inimical activity will be allowed i.e. there will be no mining, pollution creating and environment degrading activities in these areas. Yet, these very restrictions and regulations create many other opportunities. There is ample scope for a plethora of cottage industries, energy saving applications, organic agriculture, small and medium enterprises etc. Special attention has to be given to primary & secondary education, basic health care and sanitation. In fact the eco-compatible development of villages requires a much more intensive and focused effort. The Dos and Don'ts of eco-compatible development have to be followed. The people in these villages should feel privileged to live in or around the PA. But for this a strict conceptual and principled approach has to be adopted. It has to be made clear beforehand, that there are certain restrictions which will remain forever. In lieu of that there is a life-style which is not urban, yet it may be more worthy. Let there be a clear demarcation between the urban areas, the rural areas and the eco-compatible villages. In a big landscape urban areas may be there. But it is not proper to classify every urban area as developed and every rural area as backward or underdeveloped. This misclassification is in itself the root cause of many social problems. Let not the sprawling urbanization gulp down the rural areas. Let eco-compatibility have a level playing field. Let the people have choices.

In fact eco-compatibility means imbibing the spirit of the above six approaches. It means the implementation of the package of principles at the micro level. It is here that the forest deptt., the Joint Forest Management Committees and sincere wild lifers & environmentalists have complementary and challenging roles to play.

Spreading the use of alternative sources of energy and adopting soil & water conservation on a watershed basis are other areas which need to be emphasized. But if the above seven points are taken care of there may be ample optimism for the survival of tiger and natural ecosystems. These points may appear somewhat simple. But great difficulty may be encountered when it comes to actual implementation. Every sound step may antagonize the big business and corporate houses. Big business, especially if it is based on natural resources, is intrinsically inimical to natural ecosystems. Big business wants a *laissez-faire* policy. Conservation of tiger and natural ecosystems requires much of reasonable and responsible regulation. In the larger and long term interest of the society, a strong principled approach is very much required (**Fig. 7.4**). Can we pledge to call back the retreating tiger? (**Fig.7.5**)

* * *

BOX 7.1

Environmental pollution is not the only problem with pesticides. Pests develop resistance to synthetic pesticides, making it necessary to constantly develop newer and more powerful (hence more costly) pesticides. What's worse, pesticides are more damaging to friendly insects and natural predators of pests; pesticide use strengthens pests

This vicious cycle is called the pesticide treadmill. Integrated pest management (IPM) combines several approaches including pesticides. It strives to prevent or delay resistance.

Non-pesticidal management (NPM) uses techniques developed and proven under IPM, but completely does away with synthetic pesticides. So, how does it protect crops? In two ways: by promoting the sharing of know-how on crops and pests; and by utilizing locally available, low cost inputs.

Insects have a four-stage cycle, and they damage crops only in the larval stage in most cases. Effective control is that which prevents the insect reaching the larval stage. There is a range of options to do this, depending on insect behavior and crop ecology:

Deep-ploughing in summer exposes pupa of insects

Promoting natural predators and friendly insects

Light traps or bonfires attract and kill adult insects

Trap crops (like marigold or castor in cotton) attract bollworms to lay eggs on them instead of the main crop

Pheromone trap attracts male insects

Simple shaking of plants like pigeon pea helps shed bollworms

Spraying the extract of neem seed kernel (chilli-garlic paste) helps control insects at larval stage

Spraying the extract of cow dung and urine repels insects as well retards their growth

All these methods have been validated and accepted by the scientific community under IPM. Indian agricultural scientists commonly believe that farmers cannot understand these 'alternative' methods. NPM developers claim farmers can understand these approaches—and improvise them according to local needs. But only if the community strategy is not 'product-centric' (revolving around marketable commodities) but 'knowledge-centric'. So, replacing synthetic pesticides with biopesticides doesn't solve the problem; it has to be a paradigm shift. NPM developers say alternatives work only when pesticides are eschewed completely.

NPM differs slightly from organic farming, which requires that farmers do away with *all* chemical inputs. NPM doesn't require farmers to give up chemical fertilizers, though it tells them that their use makes crops more susceptible to pests. NPM is driven by economics and farmer self-reliance. In fact, it is a good entry-level exercise for organic farmers.

—(No Pesticides: Down To Earth-May 31, 2006)

* * * *

BOX 7.2

Chakora Bhaji ignites hope in their crippled lives

As per a study by the National Institute of Nutrition (ICMR), Hyderabad, the calcium in the wet leaves of *Cassia tora* is 520mg/100gm and in dry leaves it is 3200mg/100 gm. The wet leaves have 82 mg/100 gm of Vitamin C and 12.4 mg/100 gm of iron.

Dr. Tapas Chakma, Deputy Director of the Regional Medical Research Centre for Tribals, a unit of ICMR in Jabalpur says, "This vegetable is commonly found in rural areas and its consumption has been very effective in treating flourosis."

Chakma, who has been working on the causes of flourosis, since the desease came to light in 1995 maintains that about 12-20 gm of dry chakoda bhaji or 80-100 gm of green chakoda bhaji should be consumed by an adult to meet the minimum requirement of 400 mg per day. An evaluation done after nutritional supplement showed complete reversal of bone deformities due to flourosis in mild cases and partial reversal in severe cases.

—Hindustan Times, Bhopal; 22nd Feb. 2008

CHAPTER VIII
EPILOGUE

While these ideas were being penned down, reports that Panna has become almost devoid of tigers. In 2006 there was a healthy tiger population. In 2009 no tigers! It is a state of affairs which has to be sincerely pondered over. What has gone wrong in a very short period? Two investigation teams were formed—one by the Central Govt. and the other by the State Govt. The Central Govt. team has submitted its report. The State Govt. team will also be doing so very soon.

In the meantime two tigresses, one from Bandhavgarh and the other from Kanha, have been translocated to Panna. This has been done in an attempt to somewhat rectify the skewed sex ratio of tiger population in Panna. It is hoped that this will also increase the genetic diversity of tigers and the future progeny will be genetically stronger. But the critics will take this as an opportunity. It will be alleged that they had always been forecasting such a situation.

One of the two translocated tigresses was airlifted to Panna (**Fig. 8.1**). This procedure of translocation is an achievement in terms of technical inputs. With this successful operation the decision on translocation of wild animals can be taken more confidently. But the note of caution is that the aspect of translocation is only a small component of the whole gamut of ecosystem management. This skill is very much being misused in Africa where thousands of wild animals are unnecessary relocated for commercial purposes.

The African model of wildlife management is often eulogized and quoted as an example to be emulated. But in essence this is not wildlife or ecosystem management at all. It is in reality manipulation and culling of wild animal population to suit commercial ends. Mass tourism was first introduced in East Africa in the 1950s with the legalization of hunting and culling of wild game by the British colonizers who controlled Kenya and Tanzania. The need for exclusive hunting and recreational zones inaccessible to "natives" led to the creation of protected areas, national parks and game reserves. But 70% national parks and game reserves in East Africa are on pastoral land, particularly Masai land. They were robbed of huge chunks of their grazing land. The Masai Mara National Park in Kenya and Ngorongoro Conservation Area in Tanzania are both examples of ecological disaster arising from tourism. In both these areas lodges and camps have heavily deforested the small riverine forests that existed here with their huge demand for firewood for cooking and heating. The network of roads and tracks for "game drives" deteriorates the grass cover which has serious consequences for plant and animal species. Having experienced the devastating effects of tourism at the nearby Masai Mara National Park, the Masai from Loita hills, some 320 km. southwest of Nairobi, have been fighting a fierce battle to prevent the indigenously Naimina-Enkiyio or the Forest of the Lost Child, from being turned into another eco-tourism destination (Komuaro, Ole: 2008).

The situation in India is quite different. It is our constitutional duty to have compassion for living creatures. Our wild life or ecosystem management starts with this compassionate spirit. The protected areas have been created as sanctuaries for wildlife. Revenue earning by efficient management is acceptable, but business is not the primary aim. The National Forest Policy, 1988 makes it amply clear—

"The Objectives of this revised Policy cannot be achieved without the investment of financial and other resources on a substantial scale. Such investment is indeed fully justifying considering the contribution of forests in maintaining essential ecological processes and life support systems and in preserving genetic diversity. Forests should not be looked upon as a source of revenue. Forests are a renewable natural resource. They are a national asset to be protected and enhanced for the well being of the people and the Nation."

With increasing biotic pressure it may be required to fence large chunks of inviolable ecosystems. But it should be done in a manner which does not obstruct the corridor connectivity and gene pool flow. The issue of corridor connectivity has long been neglected. It needs to find its due place in every landscape planning. There are instances where wild animals have been found to traverse long distances without any apparent reason. In 1924 Maharaja of Gwalior released 3 pairs of African lions in the forests of Shivpuri. In 1929 one male lion travelled a long distance of about 300 km. to reach the Sarkoha forests of Panna. There it was shot by Maharaja Yadvendra Singh. The testimony of the event lies in the form of a lion trophy in the halls of the Panna Palace (**Fig. 8.2**). The correspondent of Nav Bharat Shri Arun Singh has done a detailed reporting on this issue.

Certain protection and biological indicators need to be evolved which may help in continuously monitoring the ecosystems and guide in the interventions required. The gist of the issue is that with certain concrete efforts, India may be a torch bearer and guide in the field of wildlife and ecosystem management. Our foundation of ecosystem management is much more principled, nobler and sustainable than the apparently appealing uncertain business models which remain cropping up in different forms. What is required is a consistent and confident adherence to certain concepts, principles and values.

While we pondered over the various issues of wildlife management, our hero Bukkha got lost somewhere. He has not been located in Panna for the last many months. No one knows what has happened to him. In his prime days he was the centre of attraction for visitors and photographers. One artist was so much fascinated that he has painted a beautiful portrait of Bukkha (**Fig. 8.3**). Let us hope that Bukkha is camping somewhere in or around the Park and he will return soon for a photo session.

The Crisis of the Tiger has brought to the fore many issues and has activated the minds of many thinkers. With prompt and proper interventions Panna may again be rich in tigers. With a quick revival, Panna may once again gain the status of the best managed Park of the country. Let all sincere well wishers of Panna join hands and bring the tigers back to Panna, so that once again it is a vibrant ecosystem, a safe sanctuary and a breeding ground for tigers.

As we complete our prayers to the Almighty, there is some good news from Panna. Two sub adult tiger cubs have been sighted near the Park. May this trend continue to a happy conclusion.

REFERENCES

Ahmad, Shahbaz. Charger, the Long Living Tiger. Allahabad: Print World, 2002.

Ali, Demisroy. Inheritance and Evolution. Ankara: Metekson Publishing Co., 1984.

Ali, Salim. The Book of Indian Birds. Mumbai: Bombay Natural History Society, 1996.

Burton, Robert. Bird Migration. London: Aurum Press Limited, 1992.

Invertebrates (Hindi). Ahmedabad: Centre for Environmental Education, 2000.

Chawdhry, P. K. Management Plan for Panna National Park. Panna: Panna Tiger Reserve, 2002.

Darwin, Charles. The Origin of Species—by means of Natural Selection or the Preservation of favored Races in the Struggle for Life. Delhi: Goyal Publishers and Distributors, 2006.

Editorial. New Delhi: The Times of India, 11th Oct. 2006.

Gadagkar, Raghvendra. Survival strategies—Cooperation and Conflict in Animal Societies. Hyderabad: University Press (India) Limited, 2005.

Guidelines for preparation of Tiger Conservation Plan—Technical Document: NTCA/01/07. New Delhi: National Tiger Conservation Authority, Ministry of Environment & Forests, 2007.

Hamlyn. Illustrated Encyclopaedia of Birds. London: Hamlyn Publishing Group Limited, 1989.

Invertebrates (Hindi). Ahmedabad: Centre for Environmental Education, 2000.

Komuaro, Ole. Voices From Africa. UN-Non_Governmental Liasion Service, 2008.

Lutz & Lux. Humanistic Economics—The New Challenge. New York: The Bookstrap Press, 1988.

Luard, C. E. Panna State Gazetteer. Indore(MP): Central India Agency, 1907.

Macdonald, David. The Velvet Claw—A Natural History of the Carnivores. London: BBC Books, 1992.

McGavin, George C. Insects and spiders. London: Dorling Kindersely Limited, 2005.

Mattison, Chris. The Encyclopaedia of Snakes. London: Blandford, 1998.

Nowell, K. and Xu Ling. Taming the tiger trade: China's markets for wild and captive tiger products since the 1993 domestic trade ban. Hong Kong: TRAFFIC East Asia, 2007.

Seidensticker, Christie & Johnson. Riding the Tiger. Cambridge: University of Cambridge, 1997.

Thapar, Valmik. Tiger—The Ultimate Guide. New Delhi: Oxford University Press, 2006.

Yahya, Harun. Miracle in the Ant. New Delhi: Goodword Books, 2002.

Yahya, Harun. Darwinism Refuted. New Delhi: Goodword Books, 2002.

INDEX

A

Adam Smith 25, 68
African model 77
All-Knowledgeable 28
All-Powerful 28
Almighty 78
Altruism 25
Arun Singh 78
Avian flu 63
Ayurveda 70

B

Banda 17
Basic needs 69,72
Bee-hive 27
Bench type topography 19
Best Managed Tourist Friendly National Park 20,61
Bhuiaonla (Phyllanthus niruri) 71
Big Bang theory 28
Boas 28
Bombay Natural History Society(BNHS) 30
Brahminy duck 21, 26
Bundelkhand 17
Bukkha 20, 30, 57, 58, 59, 62, 64, 66, 78

C

Cambian period 24
Camera trap 59, 60
captive breeding 63, 64, 65
Carnivores 29
Chakora (Cassia tora) 75
Chambal 17, 69
Champion & Seth's classification 19
Charles Darwin 23, 24

Charles Dawson 25
chausingha 30
Chhatarpur 18, 19
Cicada 22, 27
Class struggle 26
Classical economics 27
Conservation 62
Corridor connectivity 78
Cosmic egg 28
Creator 23
Crop damage 71, 72
Crop protection wall 72

D
Darwinism 25, 29
Development 64
Diclofenac 71
Dung beetle 21, 26, 27

E
Eco-compatibility 73
Eco-compatible 67
Eco-compatible development 67, 72
Ecological economics 27
Ecosystem 59, 63, 64, 65, 78
Ecosystem management 77, 78
Eco-tourism 63
Engels 26
European cuckoo 21
Evolution 22, 27, 29

F
Freud 26

G
Gandhian model of development 67
Gangau Sanctuary 19
Gene pool flow 78

Genes 24
Genetic mutation 24
Genetics 24
Gharial 69
Gond 17
Gwalior 78

H
Henry Osborn 25
Herbal healthcare 67, 70
Hinouta plateau 19
Honey bee 22, 27
Humanist economists 26

I
Irreducibly complex 24
Insect 22, 74
IUCN 68

J
Joint Forest Management Committees73

K
Kalmegh (Andrographis paniculata)71
Kardhai (Anogeissus pendula) 19
Koustubh Sharma 30
Kehrvan 17
Ken 19, 30
Ken Ghadiyal Sanctuary 18

L
Laissez-faire policy 73
Landscape 67
Landscape planning 67, 68
Livelihood 62, 64
Lokendra Singh 18

M
Madhya Pradesh 17, 60
Madla plains 19
Maharaja Chhatarsal Bundela 17
Maharaja Nirpat Singh 17
Maharaja Yadvendra Singh 78
Mammal 24
Marx 26
Masai Mara National Park 77
Mass tourism 77
Miacids 29
Microbiology 24
Migration 21
Mughals 17

N
Narmada 17
National Forest Policy 77
National Park 28
National Tiger Conservation Authority 60, 67
Natural ecosystem 21, 66, 67, 69, 73
Natural selection 22, 23
Nature 22, 24, 26, 27, 28, 70, 72
Nav Bharat 78
Nebraska man 25
Nephrotoxicant 69
Ngorongoro Conservation Area 77
Nigrani system 20
Non-pesticidal movement 70, 74

O
Organic agriculture 70, 72
Organic farming 69, 70, 74

P
Package of principles 67, 73
Panna 17, 18, 20, 60, 61, 76,78
Panna National Park 18, 19, 30, 57, 58

Panna sanctuary 18, 19
Phansia 17
Philanthropy 63
Pilocene period 25
Pollination 26, 66
Principled approach 72, 73
Protected area 21, 22, 23
Public Interest Litigation 60
Pugmark methodology 60
Python 22, 28

R
Radio collaring 57
Recycling of nutrients 27

S
Sanctuary 28
Shivpuri 78
Sri Racha Tiger Zoo 63
Struggle for existence 21, 26

T
Talgaon plateau 19
Teak 19
Territorial animals 58
Territory 58
The Wealth of Nations 26
Theory of evolution 23, 24, 25
Tiger 17, 18, 29, 30, 57, 58, 59, 60, 61, 62, 63, 64, 65, 73, 76
Tiger Conservation Plan 67
Tiger Conservation Unit(TCU) 19, 20
Tiger habitat 20, 61
Tons 17
Tourism 30
Tourism zone 57, 58, 60

Tranquilisation 57
Translocation 76
Trophy 78

U
Uttar Pradesh 17

V
Vipers 28

W
Wildlife Estimation exercise 59, 60, 61
Wildlife Institute of India 59
Wildlife Protection Act 19

Y
Yamuna 17

Z
Zero waste economy 68, 69

ENDNOTES

[1] This Chapter is based on a paper accepted for oral presentation at the International Bio-Ed Conference 2008 in France.

[2] The customary hunting by tribal had been mainly for food. With no commercial intent, it hadn't been very harmful for the ecosystem.

Made in the USA
Monee, IL
04 November 2026

67377590R00052